Thousand
Million
Billion
Trillion

Thousand Million Billion Trillion

by
Nikhil Iyengar

Copyright © 2020 Nikhil Iyengar

All rights reserved

ISBN: 979-8642810118

Contents

Part I: Distance 1
Part II: Time 9
Part III: Other Measurements 15
Part IV: Physical Objects 23
Part V: Conceptual Objects 39
Notes 49

1,000
1,000,000
1,000,000,000
1,000,000,000,000

Part I

Distance

Inches

An inch is about the width of a man's thumb, according to King David I of Scotland. In French, the words for thumb and inch are the same: *pouce*.

A **thousand** inches is about the length of a tennis court.

A **million** inches is about as long as Manhattan.

A **billion** inches is about the distance from the North Pole to the South Pole.

A **trillion** inches is about one-sixth of the distance to the sun.

Miles

A mile is about a thousand "paces" for a man, according to the ancient Romans. A "pace" is every time your left foot hits the ground when walking.

A **thousand** miles is about as long as the California coastline.

A **million** miles is the distance you would travel going to the moon and back, twice.

A **billion** miles is about the distance to Saturn when it is farthest away.

A **trillion** miles would get you just $1/25^{th}$ of the way to the nearest star, Proxima Centauri.

Light-years

A light-year is the distance that light travels in one year. A light-year is about six trillion miles.

A **thousand** light-years is about a third of the distance to the nearest black hole, V616 Mon. It is also about 1/25th of the distance to the center of the Milky Way.

A **million** light-years is about half the distance to the nearest galaxy, Andromeda.

A **billion** light-years is about one tenth of the distance to the Hercules-Corona Borealis Great Wall, the largest known cosmic structure, containing billions of galaxies.

A **trillion** light-years is larger than the width of the observable universe.

Part II

Time

Seconds

A second is about the time between beats of a human heart.

A **thousand** seconds is about the world record for a person holding their breath.

A **million** seconds is about the world record for a person going without sleep.

A **billion** seconds is about how long the average person lived at the start of the 20th century.

A **trillion** seconds ago, humans were living as hunter-gatherers. The last ice age was taking place and woolly mammoths were alive.

Years

A year is the time it takes for the Earth to revolve around the sun.

A **thousand** years ago, the Chinese had discovered gunpowder and had started using it in "fire arrows".

A **million** years ago, Homo Erectus, a close ancestor of humans, started to control fire and use stone tools.

A **billion** years ago, the first multicellular life in the form of red and green algae appeared.

A **trillion** years ago was well before the Big Bang started our universe.

Part III

Other Measures

Pounds

The pound is based on the *libra*, an ancient Roman unit of weight. The pound still uses *lb* as its abbreviation.

A **thousand** pounds is about the weight of a horse.

A **million** pounds is the weight of a typical white, fluffy cloud.

A **billion** pounds is about the weight of One World Trade Center, the tallest building in the United States.

A **trillion** pounds is about the weight of an 800 foot high hill.

Gallons

An adult has a little over a gallon of blood.

A **thousand** gallons of water would fill an aquarium the size of a minivan. A small shark could live in it.

A **million** gallons of water would fill about two Olympic sized pools.

A **billion** gallons is about the volume of a large covered sports stadium.

A **trillion** gallons is about one eighth of the capacity of Lake Mead, the largest reservoir in the United States.

Horsepower

A horse can actually produce about fifteen horsepower during a peak effort lasting a few seconds.

A **thousand** horsepower is produced by a powerful sports car.

A **million** horsepower is produced by a nuclear power plant.

A **billion** horsepower is about the power production capacity of all solar panels in the world.

A **trillion** horsepower is about the power of a hurricane.

Part IV

Physical Objects

Atoms

An atom is a chemical element that forms matter.

A **thousand** atoms is about twice the number of atoms in melittin, the major pain producing molecule in honeybee stings.

A **million** atoms make up a rhinovirus, the main cause of the common cold.

A **billion** atoms make up one fourth of the human Y chromosome, a DNA molecule which determines a person's sex.

A **trillion** atoms make up one third of a human red blood cell.

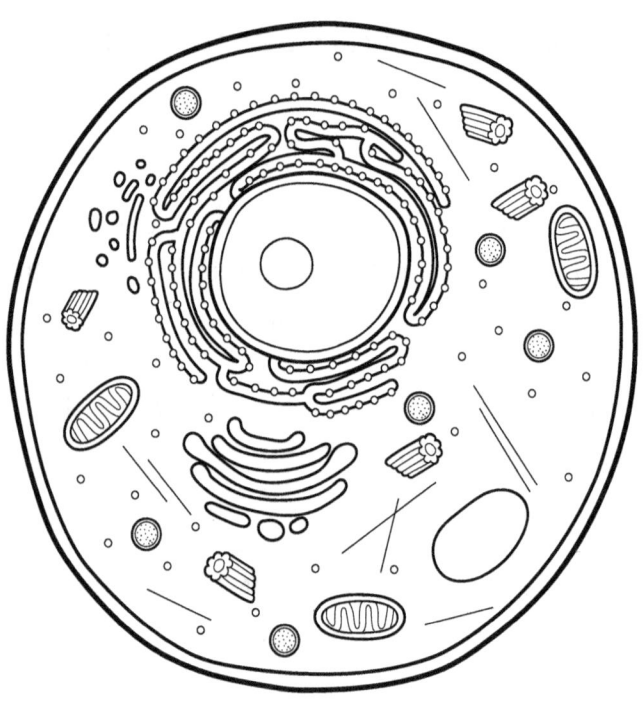

Cells

A cell is the smallest unit of life. Different kinds of cells can have very different sizes.

A **thousand** cells make up the tiny roundworm *C. elegans*, which is about one millimeter long and lives in the soil.

A **million** cells make up a small blade of grass.

A **billion** cells make up half of the muscle cells in the human heart which beat in synchrony.

A **trillion** cells make up 1/40th of the human body.

Sand

Most sand on Earth is made of quartz.

A **thousand** grains of sand can fit on a fingernail in a single layer.

A **million** grains of sand is about two tablespoons of sand.

A **billion** grains of sand would fill four basketballs, each weighing 25 pounds.

A **trillion** grains of sand would fill a ten foot by ten foot by ten foot room and weigh 50 tons, or about as much as 25 cars.

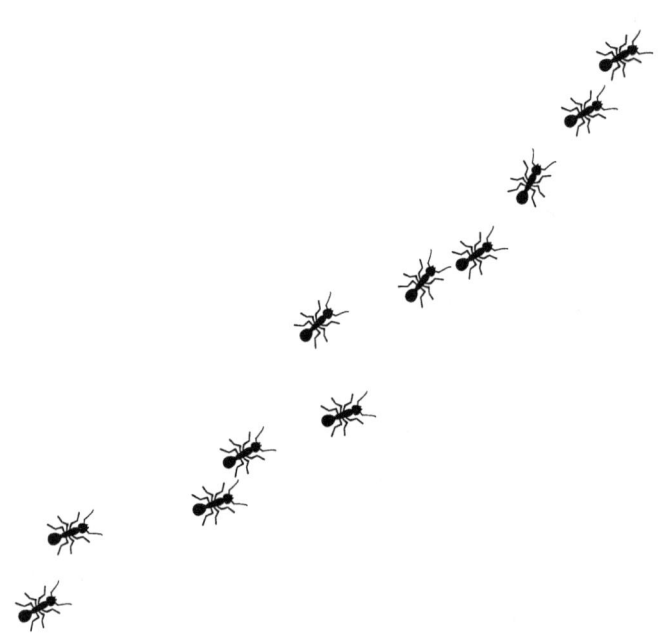

Insects

More than 90 percent of the animal life forms on Earth are insects.

A **thousand** wasps may be found in a wasp nest.

A **million** termites may be found in an underground termite colony.

A **billion** insects is about four times the number of insects per acre of land on Earth.

A **trillion** locusts were part of one of the largest locust swarms ever.

People

Humans are the only animals that cook food and wear clothes.

A **thousand** people could be on a New York City subway train during the morning rush hour.

A **million** babies are born every season in the United States.

A **billion** people live in China, the most populated country on Earth.

A **trillion** people is more than the number of people who have ever lived.

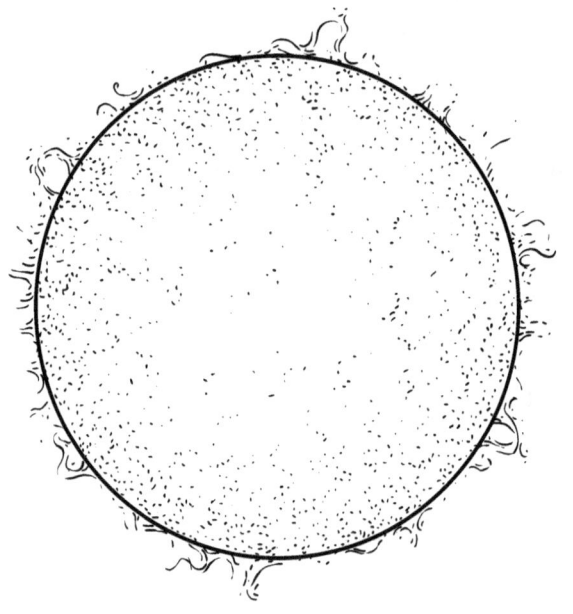

Stars

Nuclear fusion causes stars to shine.

A **thousand** stars make up the Pleiades star cluster, the most visible star cluster in the night sky.

A **million** stars make up about one tenth of the Omega Centauri globular cluster, the most massive star cluster in the Milky Way.

A **billion** stars is the number of stars we have mapped in detail.

A **trillion** stars is about triple the number of stars in the Milky Way.

Galaxies

A galaxy is a gravitationally bound system of stars, stellar remnants, interstellar gas, dust, and dark matter.

A **thousand** galaxies make up the Virgo cluster, the nearest galaxy cluster.

A **million** galaxies is about ten times the number of galaxies in the Laniakea Supercluster, which contains our Milky Way.

A **billion** galaxies make up the Hercules-Corona Borealis Great Wall, the largest known structure in the universe.

A **trillion** galaxies is about the number of galaxies in the known universe.

Part V

Conceptual Objects

Money

Only 8% of the world's money is in the form of bills and coins. The rest is recorded electronically.

A **thousand** dollars is about the cost of a high end smartphone.

A **million** dollars is about the price of a home in San Francisco.

A **billion** dollars is about twice the cost of an Airbus A380, the world's largest passenger plane.

A **trillion** dollars is about how much is spent each year on the United States military.

Words

The most common first word for babies in English speaking countries is "dada".

A **thousand** words make up the Declaration of Independence.

A **million** words make up the *Harry Potter* book series.

A **billion** words make up about one fourth of the English Wikipedia.

A **trillion** words is about half of all the words in all published books.

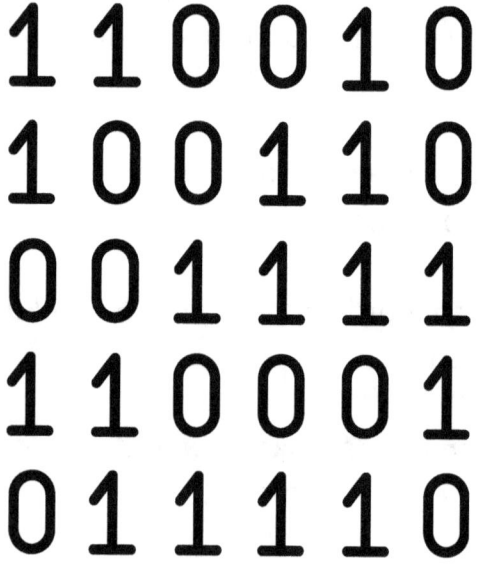

Bits

A bit is the basic unit of binary information. It can have a value of zero or one.

A **thousand** bits can store a short text message.

A **million** bits can store a photograph.

A **billion** bits can store about two hours of music.

A **trillion** bits can store about ten movies at very high quality.

Possibilities

A possibility is something that could happen.

A **thousand** possibilities cover the ways of ordering six objects in a row.

A **million** possibilities cover about half of all five card poker hands.

A **billion** possibilities cover about three times all the Powerball lottery number picks.

A **trillion** possibilities cover about half of all endings of a Connect Four game.

Notes

Inches

An inch is 2.54 centimeters.

Thousand – A thousand inches is about 83 feet. A tennis court is 78 feet long.

Million – A million inches is about 15.8 miles. Manhattan is 13.4 miles long.

Billion – A billion inches is about 15,800 miles. The distance from the North Pole to the South Pole along a line of longitude is about 12,400 miles.

Trillion – A trillion inches is about 16 million miles. The distance from the earth to the sun is about 93 million miles.

Miles

A mile is about 1.6 kilometers.

Thousand – The California coastline is 840 miles long as measured by the Congressional Research Service using large scale nautical charts.

Million – The distance to the moon is about 239,000 miles.

Billion – Earth and Saturn are about a billion miles from each other when they are on opposite sides of the sun.

Trillion – Proxima Centauri is the nearest star not including our sun. It is about 25 trillion miles away.

Light-years

Thousand – The distance to V616 Mon is about 3,500 light-years. The distance to the Galactic Center is about 27,000 light-years.

Million – The distance to Andromeda is about 2.5 million light-years.

Billion – The distance to the Hercules-Corona Borealis Great Wall is about 10 billion light-years. It has a roughly estimated 4 billion galaxies.

Trillion – The observable universe is about 93 billion light-years wide.

Seconds

Thousand – A thousand seconds is 16 minutes and 40 seconds. Branco Petrovic held his breath for 11 minutes and 54 seconds. Aleix Segura held his breath, after inhaling pure oxygen, for 24 minutes and 3 seconds.

Million – A million seconds is about 11 days and 14 hours. Verified record holder Randy Gardner stayed awake for just over 11 days.

Billion – A billion seconds is about 31 years and 8 months. Average world life expectancy at birth in 1900 was about 32 years.

Trillion – A trillion seconds is about 31,700 years. Anatomically modern humans appeared around 200,000 years ago. The last ice age was from about 115,000 to 12,000 years ago. Woolly mammoths went extinct around 4,000 years ago.

Years

Thousand – "Fire arrows" were incendiary arrows, possibly even rocket powered.

Million – Homo Erectus first appeared about 2 million years ago. Stone tools from 1.7 million years ago have been found. Archaeologists estimate that Homo Erectus started to control fire somewhere between 200 thousand to 1.4 million years ago.

Billion – Multicellular life appeared a billion years ago. Single-celled life appeared 3.5 billion years ago.

Trillion – The Big Bang was about 14 billion years ago. Some scientists say that time did not exist before the Big Bang, and some scientists say that other universes (or even the same universe) existed before the Big Bang.

Pounds

A kilogram is about 2.54 pounds. A pound can refer to a unit of force or a unit of mass. One pound-force is the force of gravity acting on one pound-mass on the surface of the earth. Here, all objects are assumed to be on or near the surface of the earth, so both units apply.

Thousand – An adult horse weighs 900 to 2000 pounds.

Million – The size of the typical cumulus cloud is about one cubic kilometer. It has about a million pounds of water spread out in tiny droplets allowing it to float in the sky.

Billion – A skyscraper weighs about 5,000 tons per story, and One WTC has 104 stories.

Trillion – Assume the hill is shaped like a cone with a height of 800 feet and a bottom radius of 2400 feet. Then the volume is 4.8 billion cubic feet. Each cubic foot of the earth's crust weighs about 200 pounds, giving a total weight of 960 billion pounds.

Gallons

A US liquid gallon is about 3.8 liters.

Thousand – A typical minivan has a volume of 140 cubic feet, which is about 1047 gallons. A brownbanded bamboo shark is 3 to 4 feet long.

Million – An Olympic sized pool has about 660 thousand gallons of water.

Billion – AT&T Stadium, used by the Dallas Cowboys, has an interior volume of 104 million cubic feet which is about 780 million gallons.

Trillion – Lake Mead, about 24 miles from Las Vegas, was formed by the Hoover Dam. It has a capacity of 32 cubic kilometers, though it is currently only 40% full.

Horsepower

One horsepower is about 746 watts.

Thousand – A Ferrari LaFerrari has 950 horsepower and costs $1 million.

Million – The R.E. Ginna plant in New York is a small nuclear plant that produces 582 megawatts.

Billion – In 2019, the installed capacity of solar panels was 647 gigawatts. The installed capacity of wind turbines was similar, at 656 gigawatts.

Trillion – A hurricane releases energy at a rate of about 600 terawatts. The vast majority of this power is in the formation of clouds and rain, while a small part is the production of wind.

Atoms

Thousand – The chemical formula of melittin is $C_{131}H_{229}N_{39}O_{31}$, giving it 430 atoms.

Million – The polio virus, which has about the same size and shape as the rhinovirus, was found to have 900 thousand atoms.

Billion – Normally, men have a Y chromosome and women do not.

Trillion – A red blood cell is one of the smaller cells in the human body. Some other cells may have 100 trillion or more atoms.

Cells

Thousand - *Caenorhabditis elegans* has exactly 959 or 1031 somatic cells depending on its sex.

Million – Grass cell dimensions range from 10 micrometers to 50 micrometers.

Billion – There are two to three billion cardiac muscle cells in the heart.

Trillion – There are about 37 trillion cells in the human body (not including non-human bacteria cells), about 84% of which are red blood cells.

Sand

For this section it is estimated that 1000 grains of sand placed in a row would be about a foot long. Then each grain would have a width of 0.305 millimeters, which corresponds to medium-fine sand. It is further estimated that a grain of sand has an approximately cubic shape and that the density of sand is 100 pounds per cubic foot.

Thousand – An average woman's index fingernail is 11 x 12 millimeters, giving it an area of 132 square millimeters. This would fit 1,420 grains of sand.

Million – A grain of sand has a volume of 0.0284 cubic millimeters. A million grains have a volume of 28.4 milliliters, or 1.9 tablespoons.

Billion – An NBA basketball has a circumference of 29.5 inches, which means a volume of about 0.25 cubic feet.

Trillion – The average car in the United States weighs about 4,000 pounds (2 tons).

Insects

Thousand – Wasp nests usually have a few hundred to a few thousand wasps depending on the type of wasp.

Million – Subterranean termite colonies are usually 600 thousand to a million termites.

Billion – It is roughly estimated that there are 10 quintillion insects on Earth. Earth has about 37 billion acres of land. This means about 270 million insects per acre, or 43 insects per square inch, on average.

Trillion – In 1875 a locust swarm formed across 198,000 square miles in the western United States. It was estimated to have 3.5 to 12.5 trillion locusts.

People

Thousand – This assumes the subway train has ten cars and each car has 100 people.

Million – About 4 million babies are born each year in the United States, so about one million per season.

Billion – China has 1.4 billion people.

Trillion – The world population is about 7.8 billion. About 100 billion people have ever lived.

Stars

Thousand – The Pleiades star cluster, with over 1,000 confirmed stars, is about 450 light-years away.

Million – Omega Centauri, with an estimated 10 million stars, is about 16,000 light-years away.

Billion – The European Space Agency has mapped over 1.7 billion stars in the Milky Way.

Trillion – The Milky Way has about 100 billion to 400 billion stars.

Galaxies

Thousand – The Virgo cluster has 1,300 to 2,000 galaxies and is about 50 million light-years away.

Million – The Laniakea Supercluster has about 100 thousand galaxies. It contains the Virgo Cluster and the Local Group, which contains the Milky Way.

Billion – The Hercules-Corona Borealis Great Wall has a roughly estimated four billion galaxies and is about 10 billion light-years away.

Trillion – The current estimate is about two trillion galaxies.

Money

Thousand – The iPhone 11 Pro currently costs $999.

Million – The median home price in San Francisco is currently $1.4 million.

Billion – An Airbus A380 costs $432 million.

Trillion – In 2019, the Department of Defense spent $686 billion. An additional $200 billion was spent on Veterans Affairs.

Words

According to most studies, "dada" is the most common first word.

Thousand – The American Declaration of Independence contains 1,337 words, not including the signers' names.

Million – The seven books in the *Harry Potter* series were published over ten years, comprising 6,095 pages with 1,084,170 words.

Billion – As of 2020, English Wikipedia has about 3.6 billion words.

Trillion – According to Google Books, there are 130 million existing books containing two trillion words.

Bits

Thousand – A thousand bits is 125 bytes. Each character in a text message takes between one and four bytes depending on the encoding.

Million – A million bits is 125 kilobytes. This amount is enough to store a JPEG image of a standard sized photo.

Billion – A billion bits is 125 megabytes. One minute of MP3 music takes about one megabyte.

Trillion – A trillion bits is 125 gigabytes. Ultra High Definition 4K video uses about seven gigabytes per hour, which allows for 18 hours of video.

Possibilities

Thousand – Six objects can be ordered in 6 * 5 * 4 * 3 * 2 * 1, or 720 ways.

Million – There are (52 * 51 * 50 * 49 * 48) / (5 * 4 * 3 * 2 * 1), or 2,598,960 possible hands. Four of these are royal flushes.

Billion – In Powerball, there are five balls chosen from 69 numbered white balls, and there is one ball chosen from 26 numbered red balls. The five white balls are chosen without replacement and their order doesn't matter. So there are (69 * 68 * 67 * 66 * 65 * 26) / (5 * 4 * 3 * 2 * 1) = 292,201,338 possibilities.

Trillion – There are 1,906,046,469,499 possible endings. Of these, 1,905,333,170,621 are positions where a player has won and 713,298,878 are positions where the players have drawn.

Thanks for reading!

If you have a minute, please review my book on Amazon or your favorite review site.

You can also contact me at tmbt.book@gmail.com.

www.ingramcontent.com/pod-product-compliance
Lightning Source LLC
Chambersburg PA
CBHW050257220526
45465CB00002B/717